European Corrosion Congress (EUROCORR 2020)

Online
7-11 September 2020

ISBN: 978-1-7138-3760-2

Printed from e-media with permission by:

Curran Associates, Inc.
57 Morehouse Lane
Red Hook, NY 12571

Some format issues inherent in the e-media version may also appear in this print version.

Printed with permission by Curran Associates, Inc. (2021)

European Federation of Corrosion (EFC) and the Corrosion Association in Turkey
c/o Ines Honndorf
DECHEMA e.V.
Theodor-Heuss-Alee 25
60486 Frankfurt am Main, Germany

Phone: +49 (0)69 7564-209
Fax: +49 (0)69 7564-418

honndorf@dechema.de

Additional copies of this publication are available from:

Curran Associates, Inc.
57 Morehouse Lane
Red Hook, NY 12571 USA
Phone: 845-758-0400
Fax: 845-758-2633
Email: curran@proceedings.com
Web: www.proceedings.com

TABLE OF CONTENTS

A universal scale inhibitor

Konstantinos DEMADIS[1], Argyro SPINTHAKI[1], Michaela KAMARATOU[1], George PETRATOS[1], Georgia SKORDALOU[1], Axel TRAMAUX[2], Ghislain DAVID[2]

[1] *University of Crete, Greece*

[2] *Institut Charles Gerhardt de Montpellier, France*

Scale inhibition is of crucial importance in harvesting geothermal energy, an attractive, renewable and sustainable energy source. The majority of the geothermal reservoir waters around the world contain a variety of anions and cations, which are prone to precipitation when subjected to vast transformations of both temperature and pressure, causing failures to many parts of the plant. It is a common phenomenon that the deposits encountered in geothermal systems are multi-component inorganic solids. They are described by the term "composite fouling". In an effort to discover a "universal scale inhibitor", we have studied the inhibition efficiency of four polymeric inhibitors that contain the functional groups polyethylene glycol and phosphonic acid to combat a multitude of scaling salts. These are: amorphous silica, magnesium silicate, aluminum silicate, iron (III) silicate, zinc sulfide, lead sulfide, iron (II) sulfide, and calcium carbonate. In this talk, we present the inhibition results for each scale separately and combined together in four brines of variable severity. Data are discussed and structure/activity relationships are drawn.

Magnesium Hydroxide Inhibition in Water Bearing Systems

Wolfgang HATER[1], Duygu DISCI-ZAYED[1], Anika STRATMANN[1]

[1] *Kurita Europe GmbH, Germany*

Scale formation is a common problem in all kinds of water bearing systems. Among dozens of scale types occurring in practice this study will focus on magnesium hydroxide scaling. Magnesium hydroxide is one of the most common Mg containing scale, which mainly occurs in sea water desalination plants. The main triggering factors in its formation are the concentration of Mg^{2+}, high pH and high temperatures.

The addition of an anti-scalant into water is state of the art to mitigate scale formation. It is important to choose the proper anti-scalant according to the local conditions in order to solve the problem successfully and economically.

In this study the inhibition of magnesium hydroxide is investigated by using antiscalants with different chemistry. For general antiscalant selection beaker tests at room temperature were conducted. Those tests provided reliable results in short time. Further trials under more practice like conditions were performed using differential scale loop (DSL) using synthetic sea water.

The purpose of this contribution is to introduce and compare test methods for magnesium hydroxide inhibition and to show the structure-inhibition efficiency relationships of selected antiscalants on scales containing magnesium hydroxide in water bearing systems.

The effect of metal hydroxides on silica scale formation

Konstantinos DEMADIS[1], Michaela KAMARATOU[1], Argyro SPINTHAKI[1]

[1] *University of Crete, Greece*

Colloidal/amorphous silica ($SiO_2 \cdot xH_2O$) is one of the most undesirable deposits in the water treatment industry. It is not as widespread as other mineral scale deposits (eg. $CaCO_3$), but its formation and deposition onto critical equipment surfaces can be catastrophic. Its control is mainly based on two approaches: (a) removal of "soluble" and colloidal silica before entering the system, and (b) use of chemical inhibitors. Nevertheless, silica scale inhibition is not a simple task. The reason lies with the chemical nature of colloidal silica. Silica is a random, three-dimensional polymer that forms by propagation of Si-O bonds, which form by polycondensation of monosilicic acid. The amorphous nature of silica precludes use of the "traditional" scale inhibition approaches (usually phosphonate additives and polyacrylate-based polymers) that are principally followed for mineral scales. Silica formation is further complicated by the presence of metal hydroxides (commonly iron, aluminum, and magnesium hydroxides). The scope of the present study is to explore the effects of the presence of such hydroxides in a process system on silica formation.

Corrosion protection of carbon steels under geothermal stresses

Stefania LIAKAKI-STAVROPOULOU[1], Konstantinos DEMADIS[1], Duygu DISCI-ZAYED[2], Juergen MATHEIS[2], Andre SCHIRMER[2], Wolfgang HATER[2]

[1] *University of Crete, Greece*

[2] *Kurita Europe GmbH, Germany*

Brines in geothermal systems influence critical surfaces in two major ways: (a) they cause corrosion phenomena, and (b) they induce severe scaling. Severe stresses are posed on metallurgies used for the construction of equipment. These include high temperatures and pressures, corrosive pH values, and high dissolved solids brines, to mention a few.

In this paper we present results that describe the corrosion behavior of carbon steel at high temperatures (~ 130 °C) and pressures at a wide range of pH values and in brine solutions of variable concentrations. The corrosion has been studied in terms of water chemistry (dissolved and total iron), and the nature of corrosion products (composition, texture and morphology of iron oxides).

Also, a series of corrosion inhibitors were evaluated for their ability to protect these carbon steel surfaces, by reducing the corrosion rate and minimizing iron dissolution. Efforts are made to correlate structural features of the inhibitors with anti-corrosion efficiency.

Corrosion inhibitor use : from laboratory testing to field application

Mioara STROE[1], Raquel ARAUJO[2], Humberto RODRIGUEZ[2], Laurent DEHAYS[2], Thiago MESQUITA[3]

[1] Total SA, Angola
[2] Total E&P Angola, Angola
[3] Total SA, France

One of the most common solution for oil and gas production is the use of carbon steel and low alloy steels for the pipeline and process equipment. In order to ensure the integrity of these equipment, the choice of this metallurgy needs to be accompanied by corrosion mitigation measures, like the use of corrosion inhibitor (CI).

The present paper discuss the selection program to which a corrosion inhibitor is submitted in order to assess its performances in term of corrosion inhibition efficiency and impact on process. The testing program is adapted to be as much as possible representative for the future field conditions. The CO2 content, temperature and produced water salinity are one of the most important parameters to consider when the anti-corrosion efficiency is evaluated. Additionally, the presence of sand and flow velocity can play an important effect in the behavior of the CI . Last but not least, the CI's tendency to emulsion stabilization and foaming is evaluated. The CI Integrity Operating Window (CI-IOW) is defined this being the allowed domain of used and the mode of treatment of the selected product.

Furthermore, in the paper an operational feedback from the site application is given, through the evaluation of the corrosion inhibitor efficiency via the monitoring technique in place.

Key words: testing program, Integrity Operating Window , sand

VIRTUAL EUROCORR

The effect of different microstructures on corrosion inhibition of high strength steels

Mehdi BEHROUZ VAZIRI[1], Anthony SOMERS[1], Bruce HINTON[1], Maria FORSYTH[1]

[1] *Deakin University, Australia*

The two most commonly used high strength steels in ship building industries are HY80 and DH36. These steels are considered as preferable alternatives to mild steels due to their excellent mechanical and welding properties. Despite the higher corrosion resistance of these steels in marine environments than that of mild steels, degradation due to corrosion still costs billions of dollars annually for marine industries. One of the most cost efficient methods to minimise this corrosion is through the use of inhibitors. It has been reported that the combination of lanthanum and cerium with carboxylates can significantly enhance corrosion resistance of mild steel in chloride solutions. The performance of such inhibitors with other steel microstructures and compositions has not been assessed. In this study, the corrosion of HY80, DH36 and AS1020 mild steel in a control solution of 0.01M NaCl with lanthanum 4-hydroxy cinnamate was investigated. Electrochemical experiments together with optical microscopy, optical profilometry and scanning electron microscopy were used to study the corrosion mechanisms over 24 hours. For each steel, a reduction in corrosion with the addition of the inhibitor was observed, however the effect of the inhibitor for each steel was markedly different. HY80 with a tempered martensitic microstructure, showed the highest corrosion resistance in the control solution, but showed the highest corrosion current with the lanthanum 4-hydroxy cinnamate inhibitor present. The DH36 steel with a ferrite and pearlite microstructure, had a moderate corrosion rate in the control solution, but the effect of the inhibitor when present was greater than for HY80. AS1020 with a similar microstructure to DH36 but with less pearlite, had the highest corrosion rate in the control solution and the lowest corrosion rate with the inhibitor present. While the inhibitor was effective in terms of reducing the corrosion current for all steels, a considerable number of small pits was observed. These results suggest that the inhibitor compound interacts with an active surface, thus having the biggest effect on the mild steel. It is hypothesised that at the early stage of immersion the inhibitor molecules bond through the carboxylate group to active anodic sites adjacent to cathodic sites, such as the high carbon phases i.e. cementite/carbide and/or inclusions such as sulphides. With time this film builds up and the La hydroxide deposits at the cathodic sites to reduce their activity. This will form a protective inhibiting film, thus mitigating corrosion during longer immersion times. Further surface analytical experiments are being used to investigate the formation and morphology of the corrosion inhibitor film on the metal surface to identify the microstructural and compositional factors that lead to the preferential attachment of the inhibitor molecule to suppress corrosion.

Exploring Green Corrosion Inhibitors using Electrochemical Noise Measurements of Corrosion in acidic conditions

Douglas MILLS[1], **Josh ZATLAND**[2], **Nicola EVERITT**[2]

[1] *University of Northampton, United Kingdom*
[2] *University of Nottingham, United Kingdom*

'Green' corrosion inhibitors derived from plant materials provide environmentally friendly alternatives to conventional corrosion inhibitors. They are also much cheaper, if using a biomass waste stream or abundant plant material. There are many examples in literature of different trials, from henna leaves to celery seeds to banana peel. Although it is known that extracts contain electron rich polar atoms such as N, O, S and P which make them potentially effective inhibitors, it is difficult to predict on a molecular basis what will work well and what will not, since many interacting factors may be at play in complementary interactions. To assist in predicting the inhibition efficiency of inhibitors under varying conditions and choosing the most effective, what is needed is a short term test which will obviate the need for tedious weight loss experiments. The Electrochemical Noise Method (ENM) uses the natural fluctuations which arise during electrochemical activity to gain information about the corrosion process. A value called Resistance Nosie (R_n) can be extracted from the data as a measure of the corrosion rate which is directly related to weight loss. Using ENM is quick and non- intrusive which makes it ideal for screening. Hence a rig has been designed which allows for measurement to be made in stirred as well as static conditions and minimises the occurrence of crevice corrosion which is a hazard for EN electrodes when trying to make a comparison with corrosion inhibition calculated using the standard weight loss after immersion method. The actual instrument being used (the ProCoMeter) is also an advance on previous EN devises being a dedicated, portable, well shielded and relatively simple to use instrument. For these preliminary trials we are exploring corrosion of mild steel in HCl at a concentration of 1 Mol per L (similar to that used in the pickling industry) in both stirred and unstirred conditions at room temperature. Results will be presented comparing EN measurements with conventional weight loss experiments for both Propargyl Alcohol (a conventional industrial corrosion inhibitor) and broccoli extract at different concentration. If this work is successful then ENM will go on and be used to optimise variables like extraction technique and concentration as well as investigating the effect of temperature.

Corrosion Inhibition and Acceleration by Rare Earth Ions in Galvanic Couples

Marco OLIVEIRA[1], Alexandre BASTOS[1], Silvar KALLIP[2], Theo HACK[3], Mikhail ZHELUDKEVICH[4], Mário FERREIRA[1]

[1] CICECO, DEMaC, University of Aveiro, Portugal

[2] Institute of Chemistry, University of Tartu, Estonia

[3] AIRBUS, Central Research & Technology, Germany

[4] Magnesium Innovation Centre–MagIC, Helmholtz-Zentrum Geesthacht (HZG), Germany

RE (rare earth) ions, particularly Ce^{3+}, have been extensively studied as corrosion inhibitor for many metals and alloys. The inhibitive action is attributed to their precipitation with OH^- ions from the oxygen reduction reaction, which blocks the cathodic sites.

This communication describes conditions where, instead of inhibition, acceleration of corrosion occurs after rare earth salts are added to solution. The work was carried out mainly on the Fe-Zn galvanic couple and with various rare earth salts, using electrochemical techniques such as galvanic current measurements, scanning vibrating electrode technique and linear sweep voltammetry (polarization curves).

The increase of the galvanic current in the couple is associated to an unexpected cathodic reaction that is observed after the addition of the salts to solution. The extra cathodic current accelerates the oxidation of zinc.[1] A hypothesis is raised that it corresponds to a pre-wave of hydrogen evolution due to the reduction of water molecules in the primary hydration sphere of RE^{3+} aqua-complexes.

This work brings new information to the electrochemical behavior of RE^{3+} ions and has a significant impact on their use to control corrosion in dissimilar metal combinations, in particular galvanized steel, which is extensively used worldwide.

[1] M. Oliveira, A.C. Bastos, S. Kallip, T. Hack, M. Zheludkevich, M.G.S. Ferreira, Corrosion inhibition and acceleration by rare earth ions in galvanic couples, Journal of The Electrochemical Society, 166 (2019) 642–648.

Environmental Friendly Film Forming Technologies

Wolfgang HATER[1], Duygu DISCI-ZAYED[1], Julia JASPER[1], André DE BACHE[1]

[1] *Kurita Europe GmbH, Germany*

Operation of water/steam cycles is threatened by corrosion unless proper conditioning measures are taken. As an alternative to traditional conditioning methods such as solid alkalization or all volatile treatment, organic cycle chemistry based on film forming molecules is being used since 1960`s.

Film forming molecules inhibit corrosion by being adsorbed on the metal surface as a thin hydrophobic film. Kurita`s Cetamine® technology uses film forming amines successfully for various applications in almost all industry sectors e. g. chemical industry, power industry, food industry, etc. Depending on plant requirements, systems operate continuously or are kept under wet or dry lay up conditions. Despite their versatile and effective corrosion protection use of amines are regulated for certain applications, especially in food industry.

With this contribution, it is intended to show the use of an environmental friendly film forming technology (Cetamine® FB) which can be used especially in sensitive applications such as direct use of utility steam in food processing/preservation/packaging.

In this contribution, corrosion tests performed in beakers and at a pilot boiler and as well as electrochemical impedance spectroscopy (EIS) results will be shown for the current and the new technology. The effective corrosion protection is successfully shown for both product ranges. Whereas Cetamine® FB has a significantly improved toxicological and ecological profile.

Electrochemical and Quantum Chemical Studies on the Inhibition of Mild Steel Corrosion by a new Diphenyl imidazole derivative

Mohamed OUBAAQA[1], Moussa OUAKKI[1], Mohamed MAATALLAH[2], Abdellah JARID[2], Mohamed EBN TOUHAMI[1]

[1] *Laboratory of Materials Engineering and Environment: Modeling and Application, Faculty of Sciences of Kénitra, Ibn Tofail University, 14000, Morocco*

[2] *Laboratory of Theoretical Chemistry, Faculty of Sciences Semlalia, Av. Moulay Abdellah, P. box 2390, Cadi Ayyad University of Marrakech, Morocco*

Abstract

The subject of this work is to study the corrosion inhibition of mild steel in 1M HCl solution by the use of a new synthesized Diphenyl Imidazol phènol (DiP). For this, purpose, potentiodynamic polarization measurements and Electrochemical Impedance Spectroscopy (EIS) were carried out. The effect of DiP on the mild steel corrosion inhibition was also studied using chemical quantum calculations. The increase of inhibitor concentration led to significant corrosion rate reduction of mild steel in molar hydrochloric acid solution, with inhibitor efficiency value reaching 97%. It has also been found that the inhibition efficiency increased with increasing inhibitor's concentration. The effect of temperature on the corrosion behavior of mild steel in 1M HCl without and with inhibitor at the optima concentration of 10^{-3}M was studied in the range of temperature 298-328 K. The polarization curves showed that the inhibitor acted as mixed type inhibitor. The Nyquist plots showed that increasing DiP concentration, polarization resistance increased and the double-layer capacitance decreased, involving increased inhibition efficiency. The adsorption of the studied DiP molecule on mild steel surface has been found to obey the Langmuir isotherm model. The adsorption parameters such as enthalpy, entropy and activation energy were determined and discussed. Furthermore, molecular modeling using Density of Functional Theory calculations has been realized to correlate quantum chemical parameters to the corrosion inhibition properties found. Surface analyses were also carried out by the mean of the SEM-EDX and FTIR to confirm inhibiting process.

Keywords: Corrosion inhibition, Mild steel, Molar Hydrochloric acid, Polarization curves, EIS, adsorption, SEM-EDX, FTIR, DFT.

VIRTUAL EUROCORR

From active dissolution to corrosion inhibition - tailoring the corrosion behavior of Cu-Zn alloy using different organic nitrogen compounds

Noemie OTT[1], Sebastian SIOL[1], Patrik SCHMUTZ[1]

[1] *Empa, Laboratory for Joining Technologies and Corrosion, Switzerland*

Copper alloys such as brass (Cu-Zn) are used in a variety of industrial applications due to their versatile physical properties and good corrosion resistance. However, the presence of ions, such as Cl^-, SO_4^{2-} and NH_4^+, common pollutants in atmosphere and natural water, drastically affects their corrosion behavior. Yet, organic derivatives of ammonia are commonly used as corrosion inhibitors for copper alloys.

Nitrogen compounds can interact with copper and/or zinc to promote or inhibit corrosion depending on their types, functional groups and side chains. Despite extensive studies, these interactions are still not fully understood from a fundamental mechanistic point of view, in particular with regards to Zn dealloying.

In this study, we investigate the effect of the following organic nitrogen compounds – different amino acids (Glycine, Cysteine, Alanine, Histidine, Proline) and alkanolamines (ethanolamine, Tris(hydroxymethyl)aminomethane (Tris)), as well as 4-(2-hydroxyethyl)-1-piperazineethanesulfonic acid (Hepes) and Benzotriazole (BTA) – on the corrosion mechanisms of a Cu-30Zn in the as-received state (cold deformed) and after heat treatment. Electrochemical impedance spectroscopy (EIS) combined with X-Ray Photoelectron Spectroscopy (XPS) was used to determine the processes happening at the alloy surface as well as the surface composition after exposure to the solutions. The derived reaction mechanisms on the Cu-30Zn alloy are discussed in comparison to reactions on pure copper and pure zinc.

The EIS spectra presented three components, which could be attributed to the different dissolution kinetics of copper and zinc. The phase shift position and relative intensities can be used as fingerprints to identify the electrochemical processes occurring on the alloy surface in presence of the different organic N-compounds. Electrochemical impedance spectroscopy therefore provides a simple and effective approach to predict the affinity of ammonia derivatives with Cu and/or Zn and estimate their inhibition efficiency.

The EIS results supported by the XPS analysis indicate that the ability of a nitrogen compound to inhibit Cu dissolution is mostly related to the side chain rather than to the type of compounds. In particular, the presence of a S atom in the side chain allows to form stable CuO or Cu(II) hydroxides but no sulfur was detected on the sample surface exposed to solution containing either Cysteine or Hepes. On the other hand, the presence of amino acids in solution promotes Zn dealloying due to the high affinity of these amines for Cu. In comparison, alkanolamines, such as Tris, lead to active dissolution while maintaining the alloy surface stoichiometry. They are therefore poor corrosion inhibitors. Of all investigated organic nitrogen compounds, BTA remains the most efficient corrosion inhibitor.

Guanine as an Efficient Environmentally Friendly Copper Corrosion Inhibitor in HCl Solution – An Investigation by Electrochemical and Surface Enhanced Raman Spectroscopy (SERS) Experiments

Merlin C E BANDEIRA[1], **Elaine Felix SILVA**[1], **Julio S. WYSARD**[2], **Oscar Rosa MATTOS**[3]

[1] *LNDC/COPPE/UFRJ, Brazil*
[2] *IQ/UFRJ, Brazil*
[3] *LNDC/POLI/UFRJ, Brazil*

The demand for efficient non-toxic corrosion inhibitors for copper and its alloys grows every year, since the finding that benzotriazole (BTAH) and its derivatives present unacceptably high toxicity towards plants and aquatic microbiota. In this context, imidazole derivatives have been proposed as alternatives to BTAH, due to their low cost and toxicity, although their inhibition efficiencies (IE) are rather low as compared to their triazole counterparts [1]. Recent results confirm that imidazole (Im) and 4-methylimidazole (4-MeIm) have low IE in both near-neutral and acidic chloride media due to formation of soluble compounds with Cu_2O in the former or weak physisorption of protonated imidazoles in the latter [2,3]. Based on these data, we have selected Guanine ($C_5H_5N_5O$ - Gua), an imidazole derivative and DNA base, as a candidate for a high performance inhibitor for copper in HCl media due to its high molecular weight, non-toxicity and multiple anchoring sites to the metal.

The interaction between Gua and the copper surface immersed in 0.1 mol L^{-1} HCl (pH = 2.0) was investigated by means of electrochemical, gravimetric and SERS experiments, at different inhibitor concentration (0.0001 to 0.001 mol L^{-1}). Cyclic voltammetry scans revealed that Gua adsorbs on the metal surface, lowering the peak currents and potentials of formation of CuCl and $CuCl_2^-$ as well as hydrogen reduction, a result that was confirmed by polarization curves. Formation of Gua-copper insoluble compound can also be inferred by a new peak at E = 0.0 V followed by a decrease of current density in the anodic branch of the curve when inhibitor concentration is $\geq$ 0.0005 mol L^{-1}. Weight-loss experiments revealed that IE increased with Gua concentration and immersion time, reaching 86% at 0.001 mol L^{-1} after 90 h of immersion. Such a result was corroborated by impedance measurements. Finally, SERS spectra revealed that Gua adsorbs on the metal both at cathodic and anodic polarizations, unlike Im and 4-MeIm [2,3]. At longer immersion times at the OCP, major loss of SERS signal is observed due to formation of white protecting film of guanine-copper complex.

References

[1] M. M. Antonijevic, S. M. Milic, M. B. Petrovic, Corros. Sci., 51 (2009) 1228

[2] E. F. Silva, M. C. E. Bandeira, W. A. Alves, O. R. Mattos, J. Electrochem. Soc., 165 (2018) C375

[3] E. F. Silva, J. S. Wysard, M. C. E. Bandeira, W. A. Alves, O. R. Mattos, J. Raman Spectrosc., 50 (2019) 1438

Inhibited alkaline electrolyte with quinoline-8-sulfonic acid and CaO complex additives for Al-air battery

Daquan ZHANG[1], Hanxue YANG[1], Xiaohui LI[1], Lixin GAO[1], Xiyang DAI[1]

[1] *Shanghai University of Electric Power, China PR*

The performance of Al-air battery is improved by complex additives quinoline-8-sulfonic acid (8-QS) and CaO in the alkaline electrolyte. The efficiency of the complex additives is investigated by the hydrogen gas evolution test and the electrochemical measurements. Spectral analyses, such as infrared spectra, UV-Vis spectra and fluorescence spectra, are used to study the action mechanism of the additives.It shows that the complex additives suppress the self-corrosion of AA5052 anode in alkaline solution, as well as improving the anode utilization efficiency from 47.0% to 81.9%. 8-QS and CaO have the synergistic effect via forming a complex film on Al alloy surface. The chelation of 8-QS with Al^{3+} retards the self-corrosion of Al-anode. Deposited $Ca(OH)_2$ in the alkaline electrolyte can promote the adsorption of 8-QS. Fluorescence emission is observed after discharging test, making it possible for on-line detection of Al–air battery.

Green inhibitor for corrosion inhibition of mild steel in saline solution

Harold Jhon AMES CANCHAYA[1], Karen Zita TUFINIO MIRANDA[1], Abel VERGARA SOTOMAYOR[1], Karín María PAUCAR CUBA[1]

[1] *Chemical Engineering and Textile Faculty - National Engineering University, Peru*

Oreocallis grandiflora is a plant which is native of Peru and Ecuador, normally its using to natural medicine. Nowadays mild steel is one of the materials more used in chemical industries, for water desalinization, petroleum refining, etc., but this material undergoes corrosion in these cases using acid solutions of HCl, HSO_4, etc. to pickling, this is the solution to clean the oxide, and because the acid attack is more intensive in presence of salts, such as NaCl, this is corrosive to mild steel, is common to use inhibitors of corrosion to protect it, this inhibitor of corrosion could be organic or inorganic synthetic which are not friendly with the environment for these reasons green inhibitor of corrosion is one way to decrease the industry pollution [1] [2].

In the following study the extract of Oreocallis grandiflora leaves works such as green inhibitor of corrosion in acid solution with different concentrations of NaCl in the range of 0.1% to 3.5% w/w and the inhibitor form 250 to 1000ppm; using the weight loss and electrochemical tests were used to determinate the inhibition efficiency.

The Oreocallis grandiflora extract was obtained from the ground leaves in absolute ethanol at room temperature using the maceration with stirring method [3]. Metallic specimens of size 4cm x 1cm x 0.1cm with a small hole near the upper edge of the specimen were used for the determination of corrosion rate by the weight loss test, these specimens were mechanically polished using 80 to 1000 grade emery paper in water and degreased with ethanol before use.

The results in weight loss test showed that with the increase of the inhibitor concentration it favors the increase of the inhibition efficiency up to a saline concentration of 1%, and at higher concentrations the efficiency declines slightly, in the case of 1000 ppm of inhibitor in 1M HCl and NaCl in the range 0.1% to 1% with 6 hours of immersion time reach efficiencies of 90% approximately in contrast the efficiency was 88% in absence of NaCl. The lineal relation of concentration/surface coverage vs. concentration, Langmuir isotherm, would indicate that mechanism is associated with the adsorption of the inhibitor extract on the metal surface.

The results obtained in the weight loss test was confirm with electrochemical impedance spectroscopy test and Tafel extrapolation.

The SEM micrographs of the steel samples after immersion in acid solution and 1% NaCl in the absence and presence of the inhibitor extract at 1000 ppm confirmed the presence of the inhibitor extract adsorbed on the metal surface.

The authors wish to acknowledge the CONCYTEC for financing the research project contract 155-2017-FONDECYT.

Chuquiraga spinosa as inhibitor eco-friendly of the mild steel protection in acid solution

Karen Zita TUFINIO MIRANDA[1], Harold Jhon AMES CANCHAYA[1], Abel Fernando VERGARA SOTOMAYOR[1], Karin María PAUCAR CUBA[1]

[1] *CHEMICAL ENGINEERING AND TEXTILE FACULTY, ENGINEERING NATIONAL UNIVERSITY, Peru*

The iron and steel industry promote 40 million employed around the world. In 2018, the consumption of china was 48.8% of world demand [1]. The steel is the most important raw material for innumerable products. Manufacturing process of steel is common use as pre-treatment HCl, H_2SO_4 solutions for removes undesirable rust and scales in pickling. Unfortunately, the acid corrodes steel needing to use inhibitors for protect [2] [3]. Different extract of plants has been used as eco-friendly corrosion inhibitors since it presents organic compounds such as phenolic, flavonoids and alkaloids than formed coordinate bonds with the metal surface due to presence of donator atoms as oxygen, nitrogen, sulphur, and selenium in inverse order of the electronegative O<N<S<Se [4]. In Peru, Chuquiraga spinosa has known as "huamanpinta" different studies showed her properties anti-inflammatory and antioxidants [5]. This project is about the capacity corrosion inhibition of ethanolic extract of Chuquiraga spinosa leaves in 1M HCl. The inhibitor efficiency evaluated in range of 250 to 750 ppm, immersion time from 1 to 12 hours and temperatures in range of 20 to 50°C.The ethanolic extract got for stirred two hours to 20°C. The characterization qualitative of leaves were phytochemical analysis where found phenols, flavonoids, alkaloids, and leucoanthocyanins.

They were made weight-loss and Spectroscopy Impedance Electrochemical (EIS) techniques. The result showed the increase of the efficiency with increasing concentration of the extract reach 81.24 % at 500 ppm. The result in electrochemical methods are similar.

The inhibitor effect is associated with the inhibitor adsorption in the mild steel surface by Langmuir isotherm. The temperature effect in the range of 20°C and 50°C showed decreasing efficiency with the increasing temperature. Coming to decrease to 67.38%.

The analysis of Scanning electron microscopy (SEM) showed the surface corroded and damage in absence of inhibitor. In addition, in presence of inhibitor the surface is less corrode because there is one film on the steel surface.

The authors wish to acknowledge the CONCYTEC for financing the research project contract 155-2017-FONDECYT

High temperature oxidation behavior of low carbon sulfur free cutting steel

Naibing LYU[1], Shujia LI[1]

[1] *Shougang Research Institute of Technology, China PR*

Low carbon sulfur free-cutting steel is mainly used for processing and manufacturing dimensional accuracy required parts. The billet scale is often removed inadequately due to improper control of the heating process, which causes the surface defects of the wire rod. The defect rate has been at the level of 3%, which affects the subsequent processing and use of the wire rod.

In this paper, the effects of oxidizing media such as sulfur and oxygen in a heated atmosphere on the high-temperature oxidation characteristics of low-carbon sulfur-containing ([S]% = 0.3% -0.5%) free-cutting steel are studied in the laboratory. The experimental atmosphere and conditions were set as follows:① 2% oxygen+argon, ②0.5% SO_2+argon, ③2% SO_2+argon, experimental temperature as 900-1100 °C and the holding time 0.5h-4h. Low-carbon sulfur-containing free-cutting steel billets were heated under the experimental atmosphere. The morphology and composition characteristics of iron oxide scale, the thermodynamic and kinetic mechanism of iron scale formation was analyzed, and a heating process to reduce the surface defects of low-carbon sulfur-containing free-cutting steel wire rod was proposed.

The main conclusions of the study are as follows:① In the sulfur-containing atmosphere, the scale structure is loose and porous , and there is a FeS-FeO liquid phase between the scale and the metal. The closer to the metal the higher the S concentration, which can confirm the characteristic of S inward diffusion. FeS-FeO liquid phase is detected under low concentration of SO_2 (0.5%), holding at 900°C for half an hour.② In the 2% oxygen atmosphere, the high-temperature oxidation kinetics compound parabola of low-carbon sulfur-containing free-cutting steels. As long as the heating temperature and the holding time are sufficient, the FeS-FeO phase is also generated in the residual oxygen atmosphere without S.③No matter what kind of atmosphere, S diffuses into the steel through the oxide layer. The segregation enthalpy of S⊿H = -190kJ / mol, which is easy to segregation, reacts with iron to form FeS, and finally establishes a Fe-S-O equilibrium system at the interface.

The experimental results are applied to the heating furnace of a rolling mill where the heating source of the furnace is changed to natural gas. When producing low-carbon sulfur-containing free-cutting steel, the heating temperature is controlled at 1170-1200°C, the heating time is shortened within 72 minutes, and the product surface defect rate is reduced from 5% to 1%.

High-temperature oxidation behavior of medium manganese light steel

Shujia LI[1], Naibing LYU[1]

[1] *Shougang Research Institute of Technology, China PR*

Medium manganese light steel is Fe-C (0.1% -0.6%)-Mn (4% -12%) series alloy materials with lighter density elements such as Al and Si. Through appropriate rolling and heat treatment processes, it obtains SIP (Shear Band Induced Plasticity) effect and MBIP (Microstrip Induced Plasticity) effect to improve the stability of austenite, ensuring the material of both high strength and high plasticity, while achieving lightweight. Medium manganese light steel is called the third-generation high-strength steel, and it is an international research hotspot for new automotive steels.

Elements have different effects on the high temperature oxidation characteristics of steel. With the increase of the Al content, the oxidation resistance of the alloy material improves; the Si element can cause a low melting point phase Fe2SiO4 between the scale and the steel, which makes the scale difficult to remove and affects the final product surface quality. In this research, the oxidation behavior of medium-manganese lightweight steel Fe-Mn-Al-Si was studied by using high temperature oxidation weight gain test under the condition of 900°C-1300°C, 2% O_2 + argon atmosphere for 30min. After the high-temperature oxidation experiment, scanning electron microscope and energy spectrometer were used to characterize the surface and cross-section morphology, the element content distribution and the phase composition of the oxide layer.

The results show that: ①if the temperature is kept below 1100 °C, the oxidized weight gain increases parabolically with time; ②if the temperature is maintained at 1150°C, the oxidized weight gain is in a logarithmic growth law; ③if the temperature is maintained above 1200°C, the oxidized weight gain increases linearly; ④the temperature rises as high as 1150°C or higher, the Fe2SiO4 and FeO-SiO2-Al2O3 phases liquefy, forming oxygen channels to accelerate oxidation, and also cause the dense oxide film to break and fall off. Finally, the outer layer of the iron sheet is clustered FeO-SiO2-Al2O3 crystals. When rolling medium-manganese light steel, in order to ensure the scale removal effect and improve the product surface quality, the temperature in the heating furnace should be controlled below 1150°C.

Corrosion Mechanism and Effect of the Microstructure on the Corrosion Properties of Cr3C2-NiCrMoNb Thermally Sprayed Coatings

Davide FANTOZZI[1], Mikko UUSITALO[1], Hanna KINNUNEN[1], Ville MATIKAINEN[1], Heli KOIVULUOTO[2], Petri VUORISTO[2]

[1] *Valmet Technologies Oy, Finland*

[2] *Tampere University, Finland*

Biomass and municipal solid waste are considered neutral and low carbon emissions fuels and, by providing 64.1% of the total gross inland energy consumption, they are the most used energy sources. During combustion these fuels release high amounts of corrosive alkali chlorides. Boiler steam temperatures of heat exchangers directly exposed to flue gases must be limited for corrosion control, and in turns energy yields are reduced. Moreover, solid particle erosion often exacerbates materials wastage in fluidised bed boilers. Thermal spray coatings such as chromium carbide coatings are currently used as an effective method to extend the lifespan of boiler tubes without altering the bulk properties of the coated component. In this work, Cr_3C_2-50NiCrMoNb coatings were deposited with HVOF and HVAF techniques to study their corrosion properties in KCl bearing environments. Thermogravimetric analyses were used for temperature screening to qualitatively identify the onset temperature for KCl induced corrosion and subsequently as quantitative measurement of corrosion degradation. The results were validated in isothermal oven tests and with an air-cooled probe in an actual eucalyptus bark-fired boiler. The laboratory results showed corrosion onset temperature to be below and above 500 °C for HVOF and HVAF coating respectively. Such results were supported by the field tests for the HVAF coating. The outcome of this study was the description of the chlorine induced high temperature corrosion mechanism applied to the complex microstructure of carbide sprayed coatings. The mechanism is greatly affected by the stabilization of the metastable phases in the microstructure of the coating.

VIRTUAL EUROCORR

Effect of H2S on high temperature corrosion of corrosion resistant alloys in a thermal cracking reactor for anthropogenic resources

Manuela NIMMERVOLL[1], Gregor MORI[1], Roland HAUBNER[2], Stefan HÖNIG[3], Wolfgang HOFER[4]

[1] *Montanistic University of Leoben, Austria*

[2] *Technical University of Vienna, Austria*

[3] *OMV Exploration & Production GmbH, Austria*

[4] *OMV Refinery & Marketing GmbH, Austria*

OMV refinery is currently establishing a new process for chemical recycling of post-consumer plastics. By use of a tube reactor a mixture of post-consumer plastics and a highly aromatic and high boiling liquid is thermally treated to generate a synthetic raw oil. This synthetic raw oil can later be used inside the refinery to process it to fuels or to monomers for polymer industry. Due to the organic liquid various concentrations of H_2S gas can be formed in the tube reactor together with HCl gas from pyrolysis of polyvinyl chloride, which is contained in the plastic feed. The reactor material shall be able to withstand these severe corrosive conditions.

In a laboratory tube furnace, process conditions of the thermal cracking reactor were simulated and H_2S content was varied between 0.02 and 2%. Investigated temperature range was between 420 and 580°C. Tested materials were 1.7386 (P9 steel, X11CrMo9-1), 1.4845 (X8CrNi25-21), and 1.4876 (Alloy 800HT, X10NiCrAlTi32-20) with widely varying nickel contents since in a previous study nickel has been identified as a very important alloying element.

Before the high temperature corrosion tests the samples were ground with 1000 grit SiC paper in order to achieve a uniform surface. After corrosion experiments the samples were analyzed by metallography and SEM/EDX. Additionally the mass loss of the samples was evaluated.

Results show that at 580°C 1.7386 (P9 steel, X11CrMo9-1) shows the highest mass loss already at low H_2S concentrations, while an increase of H_2S results in catastrophic degradation. For higher nickel containing alloys mass loss rates are substantially lower. A degradation model is presented and the results are critically discussed.

HIGH TEMPERATURE CORROSION RESISTANCE OF METALLIC MATERIALS IN HARSH CONDITIONS

Frédéric NOVELLO[1], Vincent DE NOOSE[1], Benoît DENIS[1]

[1] *CRM Group, Belgium*

Highly efficient energy recovery from renewable sources and from waste incineration causes new problems of corrosion at high temperature. A similar situation exists for new recycling processes and new energy storage units. These corrosions are generally considered to be caused by ashes or molten salts, the composition of which differs considerably from one plant to another. Therefore, for the assessment of corrosion-resistance of advanced materials, it is essential to precisely evaluate the corrosion rate under conditions close to industrial conditions. To be able to advise their customers in selecting the right material or the right protective coating, the present authors are developing testing equipment and related procedures to assess the high temperature corrosion rate of metallic materials. This paper describes the testing equipment developed so far and presents the results obtained concerning the resistance of the metallic studs that anchor ceramic tiles in a waste incinerator. Tested materials include 16Mo3 (0.3% molybdenum steel alloy), Inconel 625 (austenitic nickel-chromium-based superalloy), 304L, 316 and 310 (austenitic stainless steels). Temperatures range from 400 to 700°C.

AUTHOR INDEX

European Federation of Corrosion (EFC) and the Corrosion Association in Turkey
c/o Ines Honndorf
DECHEMA e.V.
Theodor-Heuss-Alee 25
60486 Frankfurt am Main, Germany

ISBN 978-1-7138-3760-2
9 781713 837602

CURRAN ASSOCIATES INC.
proceedings.com